William Willcocks

The Assuan Reservoir and Lake Mœris

Elibron Classics
www.elibron.com

Elibron Classics series.

ISBN 0-543-70407-6 (paperback)
ISBN 0-543-70406-8 (hardcover)

This Elibron Classics Replica Edition is an unabridged facsimile of the edition published in 1914 in Washington.

63d Congress, 2d Session } COMMITTEE ON RIVERS AND HARBORS HOUSE OF REPRESENTATIVES, U. S. { Document No. 17

THE ASSUAN RESERVOIR AND LAKE MŒRIS

ADDRESS

BY

SIR WILLIAM WILLCOCKS
K. C. M. G., M. I. C. E.
MANAGING DIRECTOR DAIRA SANIEH COMPANY, LATE DIRECTOR GENERAL OF RESERVOIRS, EGYPT

AT

A MEETING OF THE KHEDIVIAL GEOGRAPHICAL SOCIETY, CAIRO, JANUARY 16, 1904

WASHINGTON
1914

UNITED STATES OF AMERICA
GOVERNMENT PRINTING OFFICE

63D CONGRESS, 2d Session. | COMMITTEE ON RIVERS AND HARBORS, HOUSE OF REPRESENTATIVES, U. S. | DOCUMENT No. 17.

THE ASSUAN RESERVOIR AND LAKE MŒRIS.

ADDRESS BY SIR WILLIAM WILLCOCKS, K. C. M. G., M. I. C. E., MANAGING DIRECTOR DAIRA SANIEH CO., LATE DIRECTOR GENERAL OF RESERVOIRS—EGYPT, AT A MEETING OF THE KHEDIVIAL GEOGRAPHICAL SOCIETY, CAIRO, JANUARY 16, 1904.

THE ASSUAN RESERVOIR AND LAKE MŒRIS.

Of all the methods which Egypt has ever employed for the increase of her material wealth, there is only one which has never failed her. Whenever the country has turned to the Nile it has not been disappointed. It was so 4,000 years ago when the problems of water storage and flood control engaged the attention of the Pharaohs of the XIIth Dynasty. It is so to-day. The problems which the engineers of Amenemhat solved differ but little from the problems which we are called to solve to-day. In those ancient times the absence of masonry regulators made it more difficult to control floods than to provide additional water. For us the provision of additional water is more difficult than protection from floods. Both however are equally important. The country needs one as well as the other, and it is to show you how, in my opinion, both can best be obtained that I devote myself to-day.

All those dynasties which have grappled with drought and inundation, the two great enemies of their country, have insured the happiness of their people and deserved well to be remembered in history. When therefore at the beginning of the last century, Mehemet Ali merged the Egyptian question into the irrigation question he followed in the steps of the very greatest of his predecessors and laid deep the foundations of his country's prosperity. Since his day Egypt has steadily persevered in the path laid out for her by his masterful hand, and to-day we are considering the question of extending to the whole valley of the Nile, the perennial irrigation and consequent intense cultivation which he gave to part of the Delta, and of insuring for this future wealth an immunity from the terrible evils of inundation.

During the early years of my work in Egypt, I was so forcibly struck by the damage done to the perennially irrigated tracts by the oversaturation of the soil, that in the first edition of my book on "Egyptian Irrigation" I regretted that we could not go back to the basin irrigation evolved by the wisdom of the ancient Egyptians, an irrigation which in its grandeur and completeness can only be compared to an epic poem. Since then, however, I have seen the wonderful transformation of the country by twenty years of steady and persistent effort to improve the drainage. I have seen the science of manuring and rotating crops on one hand, and the practice of draining and irrigating by rotation on the other, make such rapid and simultaneous strides, that lands can now be made to produce their two and even three crops every year, and still retain their full vigour. We know that in future drainage works must precede irrigation works, and we can recommend a comprehensive scheme of perennial irrigation without any misgivings.

Egypt has a total irrigable area of 6¼ million acres. Of this area ¼ of a million acres, which are to-day inundated in flood and lie along the edge of the deserts, must continue to be inundated in flood for all time, to prevent the sands of the desert from spreading over the Nile Valley. Their value is £5,000,000. Four million acres are perennially irrigated. They have a mean value of £55 per acre, and have a total value of £220,000,000. Of the remaining two million acres, two-thirds are irrigated only in flood and one-third is not irrigated at all. These 2 million acres have a mean value of £25 per acre, and are worth £50,000,000. The land of Egypt may be considered as worth £275,000,000 to-day. If it were possible to perennially irrigate the 2 million acres which are without such irrigation, their value would be increased by £30 per acre, or by £60,000,000 in all.

The problem before us is how to provide perennial irrigation to these 2 million acres and so add £60,000,000 to the wealth of the country.

It has been calculated that each milliard of cubic metres of water stored in reservoirs situated in Egypt itself is sufficient to insure the conversion of half a million acres from flood to perennial irrigation. Egypt therefore requires reservoirs capable of storing 4 milliards of cubic metres of water.

In Mehemet Ali's time, the great preoccupation of the Government was the pressing on of the cultivation of cotton, and as this crop needed perennial irrigation, the securing of an abundant supply of water all the year round was the problem of the day.

The fame of the ancient lake Mœris had made a profound impression on the mind of Mehemet Ali, and he urged on his chief engineer the necessity of undertaking similar works. Linant Pasha first set himself to discover the site of the ancient lake, and then estimated roughly the cost of reconstructing it, but considered the cost prohibitive. He recommended Silsila as a suitable site for a weir and a canal head. The failure of the Barrage discouraged the Government from undertaking new works and the question dropped. In 1880 Count de la Motte proposed a dam at Silsila and a reservoir to the south of it.

He also proposed putting a capacious depression to the east of Kalabsha in communication with the Nile by the aid of a dam at Kalabsha. About two years later Mr. Cope Whitehouse suggested utilising the Wadi Rayan depression as a reservoir. This depression had been already mentioned by Linant Pasha in his book and located by him on his hydrological map. Financial difficulties and the supposed failure of the Barrage prevented the Egyptian Government from seriously considering the question of reservoirs for increasing the discharge of the Nile, as it had insufficient means for utilising the existing supply. The success of the Barrage repairs in 1887 gave new life to the question of reservoirs and Sir Colin Scott-Moncrieff deputed Col. Western to give shape to the suggestions made by Mr. Cope Whitehouse, to make plans of the Wady Rayan and the deserts between it and the Nile, to find out the capacity of the reservoir, and see if it could be utilised. Col. Western's report, plans, and estimates, were printed by the Egyptian Government in 1888. At the same time I was deputed to examine the other projects of Count de la Motte. In 1889 and 1891 I reported unfavourably on them, because I could find no depressions near Kalabsha to put in communication with the Nile, and could find no rock at Silsila on which to build a dam. The Bergat Takham pass was the only depression near Kalabsha which could have been used as a reservoir and it was over 100 metres above the level of the Nile flood; while both in the Silsila pass and the Silsila gate I bored for rock and was everywhere still in sand 10 metres below the level at which the existence of rock was assumed by the Count's engineers. On my report reaching Cairo, Mr. Prompt proposed using the trough of the Nile itself at Kalabsha as a reservoir in place of the depressions which did not exist. Col. Western left the country in 1890 and I became Director General of Reservoirs Studies. Mr. Prompt had supposed that rock could be met with at Kalabsha at a depth of 4 metres below low-water level. I could not find it at a depth of 26 metres. After sounding and boring at every possible site on the Nile and surveying, boring, and levelling in the deserts between Wady Halfa and the Fayoum, I submitted my report in 1894. I rejected all the existing projects for solid dams, and proposed an open dam at Assouan of a new type, which I felt sure would meet the requirements of a Nile reservoir-dam. Sir William Garstin, Sir Benjamin Baker, and Signor Torricelli approved of the site and the design. The Assouan dam is a type of work which, if successful, will revolutionise the construction of reservoirs along the great rivers of the arid and semi-arid regions of the world.

The Assuan Reservoir at its present level contains one milliard of cubic meters of water which will suffice for the conversion of half a million acres to perennial irrigation, adding £15,000,000 to the wealth of the country. But though the dam was only completed at the end of 1902, already the whole of the water has been devoted to special tracts, and the Government is reluctantly compelled to refuse all applications for water. This has been an unexpected blow to the country, which had been waiting patiently these six years for the long-promised reservoir. While the whole world is crying for that long stapled cotton which in its highest grades is the specialty of Egypt, the Egyptian landowners of 1½ millions of acres are compelled to wait perhaps for many years, while every country under the sun, which can grow cotton, is trying to acclimatize their special product, and they themselves are doomed to sit idle.

And yet the provision of the remaining three milliards of water is well within the competence of the country. I quote from the summary of the Financial situation contained in page 3 of Lord Cromer's Report for 1902:

The balance of the credit of the General Reserve Fund is £2,931,000 of which £854,000 is pledged.
The balance of the credit of the Special Reserve Fund is £1,678,000 of which £609,000 is pledged.
The value of the accumulated Economies is £4,991,000.
The Commissioners of the Public Debt hold £8,774,000.

This sum brings into the country some 3½ per cent. per annum. If half of it were spent on water storage and flood protection, it would add £45,000,000 to the wealth of the country which would bring in not 3½ per cent. on £8,774,000, but 7 per cent. on £45,000,000.

The strong men who undertook the construction of the Assouan dam will not long stand wavering and undecided when the whole country has fully made up its mind. They well know that while all this money is lying practically buried and almost barren, the Nile is pouring its surplus waters uselessly and unprofitably into the Mediterannean; surplus waters capable of being stored and converted into liquid gold.

I have just shown that the provision of three milliards of cubic metres of water is well within the competence of the finances of the country. I shall now show that the solution of the question presents no physical difficulties.

Egypt already possesses the germ of all the storage works she needs. Six years ago a few far-seeing men saw clearly what all of us understand to-day; but among the few, no man had greater faith in the future of the country than Sir Ernest Cassel. The Assouan Reservoir project had been lying buried for four years in official pigeon holes, when in 1898 Sir Ernest came forward with the funds, and with Sir John Aird & Co., as contractors, and Sir Benjamin Baker as Consulting Engineer, undertook to complete the Assouan dam and the Assiout weir by December 1903. The Egyptian Government advised by Sir William Garstin accepted his offer, and received the completed works by December 1902.

The Assouan dam is a granite structure 2,000 metres long which crosses the head of the Assouan cataract of the Nile in one continuous straight line. The top of the dam is a roadway 4 metres wide. The dam is 37 metres high at its highest point. The deepest foundation is 12 metres below the zero of the Assouan gauge on Elephantine Island, and the roadway is 25 metres above it. The dam is pierced by 140 under sluices of 7 metres by 2 metres for passing floods, and by 40 upper sluices of 3½ metres by 2 metres for passing the high level water of the reservoir. The sluices are regulated by "Stoney" gates worked by winches at the roadway level. At the west end is a navigation channel with four locks of 80 metres by 9 metres. As the dam exists at present it can hold up water to a height of 22 metres above the zero of the Assouan gauge and form a reservoir containing one milliard of cubic metres of water. The greatest head of water that can come on a single lock gate is 9 metres on the top gate when the reservoir is full and a boat is entering the second lock. A high flood on the Nile is about 12,500 cubic metres per second while the lowest summer discharge has been 200 cubic metres per second.

While the red, muddy waters of the Nile flood are pouring down the river, the whole of the sluices are open and the river discharges itself through them without parting with its silt. This is the real object of the sluices, for if the dam were solid and the river forced to flow over the top, the reservoir would soon be filled with deposit and obliterated, while Egypt, deprived of this rich mud, would be considerably the poorer. This is the great feature of the dam. While the dam holds together, the reservoir will be free of silt.

When the turbid flood has passed and the comparatively clear winter supply of the river has begun to arrive, the sluices are gradually closed and the reservoir filled. Beginning about the 1st December, the reservoir is filled in 100 days. It will ordinarily be full on the 1st March. No additional water is needed for irrigation between March 1st and May 1st as the river naturally has enough for the requirements of the area at present under crop at this season. As the area under perennial irrigation will be gradually increased, the demand for water for the new lands will begin about the 1st of April. The demand increases through May and June, and the reservoir will then be aiding the river with its supplement. If the flood is very late, water may be required from the reservoir till the 10th July; if the flood is early no water will be needed after the 20th of June, as in 1903, the first year of the reservoir. The earlier the flood the more effective the reservoir. By the time the flood has begun to get turbid, the under and upper sluices will all be open, and the muddy waters of the Nile will sweep through the dam without impediment. It will give you an idea of what this means to Egypt when I tell you that it has been calculated that a plentiful irrigation to a fallow field in flood has a manurial value of 16 shillings per acre.

The dam has worked for one year and given satisfaction. When the Nile was at is lowest last May, the natural discharge of the river, supplemented by all the subsoil infiltration water which enters the river between Assouan and the sea, was 400 cubic metres per second. The reservoir was adding 200 cubic metres per second to the supply, raising the total supply available for irrigation to 600 cubic metres per second. The reservoir was supplying one-third of the water which was being utilized in Egypt. This water will suffice for an increase to the perennially irrigated area of ½ million acres.

At the time of designing the dam, I intended that it should be of such a section that it could be raised 6 metres in height and hold up another milliard of cubic metres of water. Such an operation, if performed to-day, would mean: the whole length of the dam being raised 6 metres, the winches working the sluice gates being raised 6 metres and provided with new wire ropes, and new copings being given to the parapets. It would necessitate two more locks and three more lock gates, and nothing else. The expenditure incurred would be roughly as follows:—

Masonry in the dam 75,000 c. metres at £1.50	£120,000
Coping 6,000 c. metres at £5	30,000
Raising sluice-gate winches, 180 at £100	20,000
Masonry in the locks 100,000 c. metres at £1.50	150,000
Three new lock gates 700 tons of iron at £40	30,000
Compensations for lands drowned out	50,000
Earthwork and rock clearance, etc	30,000
	430,000
Add contingencies	70,000
Total	500,000

As the existing locks are 80 metres by 9 metres, while the largest boats in the country are 26 metres by 8 metres, and larger dahabiehs 30 metres by 6 metres, it would be possible to divide each of the two upper locks into two locks of 40 metres by 9 metres. The raising of the lock walls would be a simple matter since the existing walls are very wide at their bases and built up in steps ready to receive new masonry. If this method were adopted, the pressure of the lock gates would be greatly reduced by having six new locks instead of the four which exist to-day. If it were considered preferable to change the four existing locks into six locks of equal size the provision allowed in the estimate of cost would be more than sufficient to meet this expenditure.

This figure of £500,000 could not be exceeded. It would not only provide money for raising the dam, but ample funds for thoroughly doing up with the best cement masonry the downstream apron. Raised in this way with a liberal provision of money, the dam with its locks would be stronger than what it is to-day; while the extra milliard of cubic metres of water obtained at an expenditure of £500,000 would suffice for the irrigation of 500,000 acres of perennially irrigated land and add £15,000,000 to the wealth of Egypt.

I have frequently heard it objected to the raising of the dam by 6 metres that the theoretical pressures at the toes of the structure would exceed 5 kilograms per square centimetre, the maximum pressure proposed by International Commission. This pressure was proposed by Signor Toricelli as the structure was a new kind of dam, and he stated that until it was tested extreme caution should be adopted. The Gileppe dam in Belgium, which is 47 metres high, has a toe pressure of 10 kilograms per square centimetre. The Khamis dam in Algeria has 11 kilograms. Professor Rankin considered pressure between 7½ and 10 kilograms as quite safe for dams 48 metres high. When Sir Benjamin Baker was made consulting engineer to the Assouan dam I asked him his opinions. He said, "The structure must be nowhere under tension. I do not care a damn for the theoretical toe pressures. The actual strains to the work will come from expansion and contraction." Men like Sir Benjamin Baker are not frightened by paper figures. They have contemplated realities too long for that. No, the raised dam, if liberally financed, will be stronger than the existing work.

When the Assouan dam will have been raised, we shall be standing on the threshold of what it will be able to do. The projected Wady Rayan reservoir, or the modern Lake Mœris, will be well able to supply the two remaining milliards of cubic metres of water when working in conjunction with the Assouan Reservoir. The great weakness of this projected lake has lain in the fact that *by itself* it can give a plentiful discharge in April and May, less in June, and very little in July, and it was for this reason that in my report of 1894 to the Egyptian Government I had reluctantly to recommend that it be not carried out. But when the Assouan Reservoir is capable of supplying two milliards of cubic metres of water it will be possible to utilise the Mœris Lake to its utmost capacity. The Assouan Reservoir, being high above the level of the Nile can give its supply at the beginning or end of the summer; it can give it slowly or with a rush; while the projected Lake Mœris, being directly in communication with the Nile, and only slightly above low Nile level, its discharge would depend entirely on the difference of level between it and the Nile, and consequently as the summer advanced it would gradually fall and would not be able to give at the end of the summer a quarter of the discharge it could give at the begining.

But let us imagine that the reservoir and the lake are both completed and full of water, and that it is the first of April. Lake Mœris will be opened on to the Nile and give all the water needed in that month, while the Assouan Reservoir will be maintained at its full level. In May, Lake Mœris will give nearly the whole supply and the reservoir will give a little. In June the lake will give little and the reservoir much; while in July the lake will give practically nothing and the reservoir the whole supply. Working together in this harmonious manner, the reservoir and the lake, which are the true complements of each other, will easily provide the whole of the water needed for Egypt.

The Wady Rayan is a depression in the deserts to the south of the Fayoum and separated from the Fayoum by a limestone ridge. In 1888 Col. Western recommended it very strongly as a reservoir. In this he was supported by Col. Ross, the first Inspector General of Irrigation. On Col. Western's leaving Egypt, the study was entrusted to me, and Messrs. Hewat and Clifton deputed to make a final project. To Mr. Clifton, who is kindly pointing for me, I owe the longitudinal section of the Wady Rayan. To Mr. Hewat I owe the plan. Mr. Letchford has very kindly provided the excellent drawings of the Assouan dam and Lake Mœris. The Lake Mœris project, with its plans and estimates, was published by the Egyptian Government in 1894. As I said before, I was reluctantly compelled to reject its adoption owing to the one radical defect already described. That defect will have been completely removed by the completion of the Assouan Reservoir, and I stand here to-day as a strong advocate of the restoration of Lake Mœris.

The question of Lake Mœris has interested the world for centuries. For the ancients it was one of the world's seven wonders. Sir Hanbury Brown, in his book on the "Fayoum and Lake Mœris," has collected all the information available about the lake, and after a thorough examination of the question has declared in favour of the Wady Rayan being converted into a modern Lake Mœris.

Herodotus, writing about B. C. 450, was the first to describe the lake:

> Now the labyrinth being such as I have described, the lake named that of Mœris, causes still greater astonishment, on the bank of which the Labyrinth was built.
>
> The water in the lake is not derived from local sources, for the earth in that part is excessively dry and waterless, but it is brought in from the Nile by a canal. It takes six months filling and six months flowing back. During the six months of the return flow, it yields a talent of silver every day to the treasury, and during the flow twenty minæ for the fish.

Strabo, writing in B. C. 20, remarks:

> It has also a remarkable lake called the Lake of Mœris, large enough to be called a sea, and resembling the open sea in colour.
>
> Thus the Lake of Mœris is, from its size and depth, capable of receiving the overflow of the Nile at its rising, and preventing the flooding of houses and gardens; when the river falls, the lake again discharges the water by a canal at both mouths, and it is available for irrigation. There are regulators at both ends for controllmg the inflow and outflow.

Diodorus Siculus, writing at the same time, says:

King Mœris dug a lake which is amazingly useful and incredibly large. For as the rising of the Nile is irregular, and the fertility of the country depends on its uniformity, he dug the lake for the reception of the superfluous water, and he constructed a canal from the river to the lake 80 furlongs in length and 300 feet in breadth. Through this he admitted or let out the water as required.

At one time there was much discussion as to what was the Lake Mœris, but since the publication of Sir Hanbury's book there can be but one opinion. The lake covered the whole of the modern Fayoum below the level of the contour which is 22½ metres above mean sea level. The common Nile shells are to be met in myriads at any point on this contour round the Fayoum that one cares to look for them. The ordinary high flood level of Kushesha basin to-day is 26½ metres above mean sea. In Amenemhat's time, which was 4,000 years ago, the level was 4 metres lower, or at 22½ metres above mean sea. This was the highest possible level the lake could have attained in his day. In the course of time the level of the Nile valley rose by about 10 centimetres per century, but the frequent occasions on which the canal was kept closed during poor and low floods gradually silted up the channel and made it less capacious. As there are no Nile shells above the contour of 22½ metres above mean sea (except a few on the south side of the lake which have evidently been blown up by the northwest winds in sand drifts) it is evident that the gradual silting up of the channel more than kept pace with the rising level of the Nile. Eventually the silting up exceeded the rise, and that at an accelerated rate, the canal became weaker and weaker, and the Fayoum province gradually occupied the site of the lake. Lake Mœris had lasted over 2,000 years.

The connection between the Nile and the germ of the future Lake Mœris was in existence in King Menes's time, as I have been informed by Professor Sayce, but it was King Amenemhat of the XIIth dynasty who widened and deepened the canal, cleared away the rocky barriers, and converted the trifling lake of King Menes's time into the mightly inland sea which controlled the highest floods of the Nile. Those ancient Pharaohs were giants in hydraulic engineering. They were, moreover, as wise as they were courageous.

Sir Hanbury Brown has well described the action of the lake. It had a surface of 2,500 square kilometres, and being drained back into the Nile and kept at a low level it was able to take from a very high flood 20 milliards of cubic metres of water. It was quite capable of reducing a very high flood to moderate dimensions; and if injudiciously or maliciously opened in a low flood, it was capable of depriving Lower Egypt of any flood irrigation at all; and mind in those days they had practically no irrigation except flood irrigation.

I have often said that Mr. Cope Whitehouse was right when he insisted that the Ha-Uar of the Hyksos was the modern Hawara where stands the pyramid of the Labyrinth, and where were the Labyrinth and the great regulators of entry and exit for the Lake of Mœris. The two regulators, I take it, were two works parallel to each other closing the depression which connected the Nile Valley with Lake Mœris. The upper one was the existing Lahoun bank, with a pyramid at its northern extremity; on this bank to-day stand the villages of Hawara Eglan and Lahoun ("Lo-hunt," the Barrage). The other was a broad spill channel cut out of the living rock to a suitable level for passing ordinary floods where the Bahr Youssef is to-day, and in continuation of it a massive earthen dyke across the head of the El Bats ravine, which was cut in dangerously high floods. On the line of this second regulator is the existing village of Hawara el Makta or Ha-Uar of the breach; the Hawara pyramid or pyramid of Ha-Uar stands at its northern extremity. Between the pyramid and the great dyke was the labyrinth ("Lape-ro-hunt," the Barrage temple), in all probability a maze of outworks and barracks, temples and palaces, so constructed that no one from the mainland could approach the dyke. The other end of the great dyke was Hawara Makta, or Ha-Uar, which was practically a fortified island, surrounded by water. The two regulators or barrages were 10 kilometres apart. Owing to the difficulty of providing stone rubble pitching in sufficient quantities in an age when blasting powder was unknown, the reconstruction of the dyke after the passing of heavy floods entailed an expense which even an Egyptian Pharaoh considered excessive.

This Ha-Uar was the true key of Lower Egypt. Egyptologists say that Ha-Uar stood on Lake Menzaleh, somewhere between the ancient Tanis or Zoan and the modern Port Said, because Avaris is described by Manetho as a town on the sea shore. Manetho lived 1,300 years after the Hyksos. The Avaris mentioned by him has apparently left no traces of its existence in land as flat as the palm of one's hand, and yet Ha-Uar was the key of Egypt. The point where the Nile flood could be controlled before it entered Lower Egypt was surely Ha-Uar, the modern Hawara of the breach, situated on the great lake or sea of Mœris, the key which opened and shut the Nile, and, in the language of the ancients, the Nile was Egypt and Egypt was the Nile.

With Ha-Uar situated at Hawara, the site of the great Barrage of Lake Mœris, the history of Joseph's famine becomes quite intelligible. We are lifted out of a world of wonders into one of possible realities.

History tells us that Joseph arrived in Egypt late in the time of the Hyksos, who ruled Lower Egypt, while Theban dynasties ruled Upper Egypt. Between the two crowns there was an unending war. At one time the Hyksos held the country as far south as Thebes. As years rolled on the fortune of war went gradually against Lower Egypt and Upper Egypt kings won their way down the Nile valley, and about the time that Joseph arrived they might have been nearing Ha-Uar, Hawara, the regulator of Lake Mœris, and the true Southern frontier of Lower Egypt.

The very natural anxiety about the loss of the stronghold and the dire consequences which would ensue must have made Pharaoh dream repeatedly of fat and lean kine, of full ears and ears blasted with the east wind. Joseph, a thoroughly capable and shrewd man as well as a God-fearing one, while lying in prison for many years, would have learnt from his fellow-prisoners, many of them captives from Upper Egypt, that the aim of the Theban kings was the construction of a fleet and the capture of Ha-Uar. He took in the situation, and when he stood before Pharaoh, boldly told the king to put away his flattering advisers and to realise the fact that Upper Egypt was preparing a strong fleet and that when it was ready Ha-Uar might fall. The country had some years of grace, and he advised that it should spend the time in storing corn and provisions for the coming years of drought which would follow the loss of the great regulating dyke. The collection of corn was set on foot on a gigantic scale. Ha-Uar fell into the hands of the Theban kings and with it the Barrage; the Nile was opened to Lake Mœris and failed to overflow its banks in Lower Egypt. The predicted famine came. This famine was doubtless felt not only in Lower Egypt but in that part of Upper Egypt immediately to the south and north of the Lake Mœris canal, such as the island Nome and the Nome of Memphis, which were affected by the fall of the river level. Rook tombs at El kab, to an irrigation engineer the most interesting monuments in Egypt, commemorate events contemporary with these times; one tells of the deeds of a sea captain of the Theban kings who greatly distinguished himself in the capture and recapture of Ha-Uar, which preceded the fall of the Hyksos. It is worthy of note that nearly all the fighting at Ha-Uar was by water, as would naturally be the case at a fortress which was to all practical purposes an island on the edge of Lake Mœris. The other notes the fact that during the continuous years of famine which occurred at the time, the occupants did much to relieve the distress of the city. These were the same years of famine as those of Joseph. Brugsch remarks:

Since Baba lived and worked under the native king Sequenen-Ra-Taa III, about the same time during which Joseph exercised his office under one of the Hyksos kings, there remains but one fair inference: the many years of famine of Baba are the seven years of famine of Joseph's Pharaoh.

It may be that during some of these years the Nile was experiencing a series of low years such as we have had since 1899. In this case the famine in Egypt, aggravated by the opening of the Lake Mœris dyke, must have been severe indeed. "The famine was sore in the land."

Making a final effort, the Hyksos king retook Ha-Uar, and closed the dyke. The Nile flood rose to its ordinary level, and the land which had long lain fallow brought forth by handfuls. In subsequent years Ha-Uar was retaken by the Theban kings. The memory of one terrible and prolonged famine sufficed to take heart out of the people, the Hyksos fled from Egypt, and the Theban kings ruled over the whole country. There is nothing marvellous or out of the ordinary in all these striking events. Of the new Theban ruler it was said that there arose a king who knew not Joseph.

To confirm what I have said I quote from page 96 of "Egypt under the Pharaohs," by Brugsch Pasha:

In this same Sethroite Nome lay, without doubt, the town of Ha-Uar, the house of the leg, the Avaris of Manetho, a town which in Hyksos days became so celebrated. It was on the east of the Pelusiac arm of the Nile *and appears to have been connected with the river by a canal.* The gradual silting up of the bed of the river has made the situation of the towns on its banks so difficult to determine that there is scarcely a hope of finding again the site of the lost city of Avaris. But that *Ha-Uar* must in any case be sought *in the neighborhood of a lake* is evident from the inscription in the tomb of Aahames, the navigator, at El Kab, who relates how he was present when the Egyptian fleet was fighting the foreign enemies on *the waters of Pa-Zetku, near the town of Ha-Uar.* This name also, in spite of an Egyptian article Pa placed before it, has a Semitic appearance, and is doubtless connected with corresponding roots of that language.

Now the modern Hawara Eglan is situated on the dyke at Lahoun across the canal connecting Lake Mœris with the Nile, and the modern Hawara el Makta, or Hawara of the breach, is on the second dyke at the Labyrinth whose traces are to-day visible in the vicinity of the village. Hawara, the lake, and the connecting canal, all answer to the description of Ha-Uar given above. There is no necessity to get rid of the Egyptian article Pa before Pa-Zetku, because the Fayoum itself is simply Pa-ium, the lake country. The very name Ha-Uar, the house of the leg, is so appropriate to the modern Hawara on the canal joining the lake to the Nile. The statement that the silting up of the bed has made it difficult to find traces of important towns is difficult to understand, for there has been far less silting up near the sea at Tanis or Zoan than at Memphis or Cairo. True there is no sea near Hawara to-day, but in the ancient days, the Fayoum, except a narrow strip of land round Medinet el Fayoum as a centre, was a huge inland sea 2,500 square kilometres in extent, and Hawara stands on the canal joining this sea to the Nile.

Brugsch thus translates the inscription:

Pages 112, 113; from the tomb at El Kab of Aahamas, the son of Abana-Baba:

> They beseiged the town of Ha-Uar. My duty was to be valiant on foot before his Majesty. They fought by water on the Lake Pa-Zetku of Ha-Uar.
>
> After that there was a new battle at that place and I fought again hand to hand.
>
> And they fought at the place Takena at the South of the city Ha-Uar. I plunged into the water. They took Ha-Uar.

To my mind there is no doubt that Ha-Uar is Hawara, and Pa-Zetku is the lake on which Hawara stood, the ancient Lake Mœris. The name Mœris was given to the lake by the Greeks a thousand years afterwards.

We may now leave the Lake Mœris of King Amenemhat; it is the province of the Fayoum with a cultivable area of 400,000 acres and a value of over £80,000,000. We come to the projected Lake Mœris of the Wady Rayan. The Wady Rayan, as already stated, is a depression in the Lybian hills immediately south of the Fayoum. It has, at a level of about 29 metres above the sea, a surface of 700 square kilometres, or about one quarter the area of the ancient lake. Like the ancient lake, the lowest point of the Wady is 41 metres below sea level. When filled with water the greatest depth will be 70 metres. The uppermost four or five metres only will be utilised annually, or some 3 milliards of cubic metres of water out of a total volume of 20 milliards. Just as the great size of the ancient lake was of inestimable value to a work whose principal use lay in moderating high floods, so the smaller area of the modern lake will render it far more useful as a work for feeding the low Nile. This lake, as I shall presently show, will render no mean aid in time of dangerous floods, but its main use will be the provision of water in summer. It will supply the two milliards which are needed to convert the whole of Egypt from basin to perennial irrigation.

I have already explained how the lake will work in conjunction with the Assouan Reservoir. It remains to explain how it will be fed and how it will discharge its water into the Nile, what the project will cost, and how many years the lake will take to be filled and to be ready for aiding the low supplies of the Nile. In 1894, when my last calculations were made, I had a great advantage over Col. Western in that I could assume that the Assiout weir across the Nile would be in existence and capable of aiding the Lake Mœris Canal. I now have a very much greater advantage than I had in 1894. Thanks to Sir Hanbury Brown and Mr. Webb, we can assume that both the Assiut and Barrage weirs can be used not only in summer but in flood and winter. Just as the existence of the raised Assouan Reservoir will have rendered the emptying of the lake a simple matter, so the power of using both weirs in flood will have rendered the question of the filling of the lake an easy problem.

The lake can be used as a reservoir holding three milliards of cubic metres of water or two milliards of cubic metres. When containing three milliards of cubic metres of water capable of being utilized, it will have a surface area of 700 square kilomètres, a volume of 20 milliards of cubic metres of water, and a maximum depth of 70 metres. The water surface will be 29 metres above sea, and annually the 3 milliards of cubic metres of water contained between that surface and a surface of 4½ metres lower will be capable of being discharged into the Nile. When containing 2 milliards of cubic metres capable of being utilised, the water contained between a surface 27 metres above mean sea and a surface 3 metres lower will be capable of being discharged annually into the Nile by gravity. If more is needed than this from the lower lake it will be possible to discharge another milliard at a rate of 200 cubic metres per second by means of powerful pumps. Such pumps would cost £250,000 to erect, and £35,000 per annum to work. Twelve years ago such pumps were unknown in the world.

As projected for the larger quantity of 3 milliards of cubic metres of water, the lake will have a feeder canal from the Nile north of Sharaouna to the edge of the deserts north of the Mazura regulator on the Yusufi canal, and then along the edge of the deserts to the point where the lake canal begins to traverse the deserts. This canal will be 30 kilometres long and will insure the lake being filled annually to a height of 31 metres above sea level if necessary; a height of 29 metres above sea level will only be necessary. In a very high flood, however, when there is an urgent necessity this canal will be capable of raising the lake to 31 metres above sea level and relieving the Nile of 1000 cubic metres per second for 50 days or reducing its flood level by 30 centimetres for that period of time.

The main lake canal will traverse the deserts opposite Mayana and, crossing the Liernur and Masaigega Wadies, will enter the Wady Rayan at its easternmost point. This same canal from the Wady Rayan in its course through the deserts will be the outlet canal from the lake into the Nile. From its point of junction with the feeder canal it will cross the cultivated land and enter the Nile just south of Beni Souef. The line through the deserts will have a length of 16 kilometres before it reaches the low-lying Wady Liernur. The length of the outlet canal between the edge of the desert and the Nile south of Beni Souef will be 20 kilometres.

Both canals will have bed-widths of 40 metres, depth of water of 10 and 7 metres, regulators on the Nile, syphons for the Ibrahimieh and Yusufi canals, and a regulator at the point of junction of the inlet and outlet canals. This point will correspond to the Labyrinth and the regulators of Hawara or Ha-Uar of the ancient Lake Mœris. The two narrow passes which at a height of 25 metres above sea level connect the Wady Rayan with the Fayoum will have masonry retaining walls built on their rock sills.

The project will cost £2,600,000, detailed as follows:

	Cubic Metres.		
Earthwork	16.500,000	@ £0.030=	£495,000
Marlwork	7,000,000	@ 0.050=	380,000
Limestone rock cutting	8.400,000	@ 0.100=	840,000
Sand	600.000	@ 0.010=	6,000
Masonry work			375,000
Closing depressions			10,000
Land			190,000
Total			2,296,000
Contingencies			304,000
Grand total			2,600,000

The lake will provide 3 milliards of cubic metres of water working in conjunction with the raised Assouan Reservoir, as follows:—

Months.	Discharge in cubic metres per second.			
	In the Nile at Assouan and from infiltration.	Assouan Reservoir.	Lake Mœris.	Total.
April	550	0	350	900
May	500	100	450	1,050
June	400	450	250	1,100
1-10 July	450	550	100	1,100

The above will suffice, and suffice well, for every acre of land cultivable and reclaimable in Egypt.

If it is considered advisable not to construct the larger project at once, the feeder canal may be left out. In that case the outlet canal would be used to feed the reservoir during the four years of filling, and after the reservoir was full the Yusufi canal would be called upon to provide the supply of a mean discharge of 230 cubic metres per second from the 15th October to the end of February needed to fill the reservoir. The canal between the desert and the edge of the Nile would be deepened and become permanently the outlet canal.

I have been reminded by Mr. J. S. Beresford, C. I. E., formerly Inspector General of Irrigation to the Government of India, and a believer in the Lake Mœris project, that if a plentiful supply of water in the Yusufi canal could be assured in the winter, it would be possible to utilize the outlet canal as a feeder canal in flood from the Nile, and that even after the lake was full and beginning to discharge back into the Nile. It would be possible to use the Yusufi canal water to flush the canal and sweep away annually the flood silt deposits and so ensure the canal being ready to act as an outlet canal in April. This suggestion is a truly good one. Acting on it we could dispense with the proposed separate feeder canal, and so make an economy of £600,000. We might, if we preferred, make no economy, but greatly increase the size and capacity of the canal.

This latter project will cost £2,000,000, detailed as follows:

	Cubic Metres.			
Earthwork	7,500,000	@	£0.030=	£225,000
Marlwork	7,600,000	@	0.050=	380,000
Limestone work	8,400,000	@	0.100=	840,000
Sand	600.000	@	0.010=	6,000
Masonry works				225,000
Closing depressions				10,000
Land				90,000
Total				1,776.000
Contingencies				224,000
Grand total				2,000,000

The rates I have allowed for the excavation work are considered too low by some critics. If the earthwork in the Nile Valley had to be excavated within 30-day rotations as on the running canals, I should be the first to agree; but the work will last three years and the contractors will be able to concentrate all the spare labour of the country on the works when demand for labour is slack, and in this way the rate of P. T. 3 per cubic metre which I have allowed will be found to be liberal. In the hill of salted marl it will be possible to employ the American system of excavating by the aid of water issuing from nozzles under pressure. By this method it will be possible to do such work at P. T. 2 and P. T. 3 per cubic metre as it is done in America. I have allowed P. T. 5 per cubic metre. To this hydraulic pressure work the salted marls will be specially suited, and indeed the recollection of the ease with which Amenemhat dug his canal through this very material lasted long in the memory of Egyptians. Some 1,600 years after the canal was excavated, Herodotus was informed that the excavated material was thrown into the canal and transported by the running water. A 12-inch pump on the Yusufi canal lifting water on to the top of the hill, a number of spade men helping the water as it coursed down the hill and leading the liquid mud along wooden troughs into side ravines and depressions, and a steep slope on the western half of the hill where the rock had been blasted away would soon remove all the material required at a very low cost. I have allowed P. T. 10 per cubic metre for the soft limestone. Here it will be easy to work on a vertical face of some 7 metres, blast out the rock, carry it away on four lines of railway running down hill, deposit the rubble on the desert; and as each 7 metres depth is completed, to begin the next 7 metres in depth in the same way.

In my 1894 Report I had anticipated difficulties with the canal running through the salted marl. Since then I have thoroughly inspected the ravines in the Fayoum and seen the El-Bats ravine where it cuts through many kilometres of this very salted marl. The sides are absolutely vertical and deposits of mud and self-sown tamarisk bushes protect the vertical sides at places where the running water is nearly touching the marl. Such natural protection will be far superior to the masonry lining I proposed and far more effective. It will moreover cost nothing.

The lake will provide 2 milliards of cubic metres of water, working in conjunction with the raised Assouan Reservoir, as follows:

Months.	Discharge in cubic metres per second.			
	In the Nile at Assouan and from infiltration.	Assouan Reservoir.	Lake Mœris.	Total.
April	550	0	350	900
May	500	200	250	950
June	400	430	120	950
1–10 July	450	500	50	1,000

The large project will take 3½ years to carry out and the canal will fill the reservoir in 3½ years. The smaller project will take 3 years to carry out and fill the reservoir in 4 years. Utilising the Assiout and Delta weirs, it will be possible to ensure enough of water to Egypt to complete the perennial irrigation of the whole country within 7 years of the commencement of the work.

There is no more complete and perfect project before the world to-day than this combined reservoir and lake project. Look at it from its utilitarian side, it promises you £45,000,000 for an expenditure of £500,000 on the Assouan dam and £2,000,000 on Lake Mœris, or a total expenditure of £2,500,000. Look at it from its world side, it provides an acre of first class cotton for every acre irrigated perennially. Viewed from its humanitarian side it is equally attractive. To the imagination it affords one of the most fascinating views which the mind of man can conceive. It is now 4,000 years since Amenemhat undertook and carried to a successful issue his giant project of the ancient Lake Mœris. To-day, in our modern world of electricity and wonders which no Pharaoh ever conceived, in this ancient science of irrigation we are again undertaking a work not only similar to what the greatest of the Pharaohs executed, but within a stone's throw of the very same locality. Moreover, we are proposing works which are wonderfully similar to the works carried out 4,000 years ago. The area enclosed between the Nile on one side, the inlet canal on the south and west, and the outlet canal on the north was called the Island Nome of Egypt, with Herakliopolis as its capital. The same Island Nome will be cut out of the country by the works proposed to-day. There will be this exception; just as the modern Lake Mœris lies to the south of the ancient lake, so the southern boundary of the ancient island will be the northern boundary of the new one. As time goes on, the Wady Liernur and the Wady Masaigega will have their sides filled with deposit and the lake canal will flow through them as a true canal. If more water has been obtained from elsewhere it will be possible to cultivate all the lands, and as the land gradually extends into the Wady Rayan, to add 20,000 acres at first and more each year to the cultivated land of Egypt. Such land will be worth £50 per acre when covered with Nile deposit, or each 20,000 will be worth £1,000,000. If it is considered more necessary to conserve the water it will be possible to allow the level of the water in the lake to rise gradually from a full level of 29 metres above sea to 30 and 31 metres; or, seeing that the perennial irrigation of the Fayoum and the whole Nile Valley from Deirut (60 kilometres north of Assiut) to the Barrage opposite Cairo, by means of two long canals fed by one single head, will prove very difficult as the irrigation developes, I feel confident that a second weir on the Nile near the southern boundary of the Beni Souef province will become necessary. Such a weir, placed below the feeder canal head, would enable the Nile to feed the Yusufi canal downstream of the Mazura regulator, would put the irrigation of the Fayoum and Ghizeh provinces on stable footing, permit of the high-level lake being completely filled every year with clear winter water if it were judged necessary, and would allow of a canal irrigating the whole right bank of the Nile in Gizeh and the deserts north of Cairo. Such a weir would be cheaply constructed. The lock could be built on the living rock while the stone for the masonry work and pitching could be blasted from the line of the right bank canal skirting the Nile.

Before leaving this question of the re-establishment of the Lake Mœris I cannot leave untouched another possible great enterprise of King Amenemhat, the constructor of the ancient lake. At Semneh, in the second cataract, are the rocks where Leipsius discovered Nile gauges cut by the same Pharaoh over 4,000 years ago. The flood level recorded there is 8 metres higher than any flood of to-day. As the Nile at Semneh could be very easily barred by a dam, it struck me when I was there in 1892 (escorted by 150 of the Egyptian Camel Corps supplied by General Wodehouse, for the Derwishes then held the 2nd Cataract) that probably Amenemhat had tried to bar the river with a dam in the hope of creating a reservoir, and his successors had had to give up the work. The Nile in the course of centuries has regained its ancient channel.

It will readily be understood that in allowing for a provision of 4 milliards of cubic metres of water I have provided a sufficiency of water to ensure an abundant harvest under the most unfavourable circumstances. We may therefore count on being able to dispose of a far more liberal supply in every other year. The water supplied by the reservoir and the lake will not only add 1½ million of acres of perennial irrigation to the land of Egypt, but it will provide more liberally for the wants of the existing perennially irrigated area than can possibly be done to-day.

If we wish not only to irrigate the whole of Egypt but to keep the main branches of the Nile well supplied with water through the twelve months of the year, we must regulate the supply issuing from the vast lakes which constitute the sources of the Nile; we must ensure its passage through the great swamp regions; allow it of its superabundance to bless all the arid lands between the 10th and 24th parallels of Latitude; and finally see it enter Egypt a much diminished but still powerful stream of water. Writing to "The Engineer" at the time of the opening of the Assouan dam (vide page 558 of "The Engineer" of 1902) I thus expressed my opinion on this point:

It has, however, been the good fortune of Lord Cromer to lay the foundation-stone of that series of works which is to usher in an era of prosperity such as we read of in the time of the great Pharaohs of the twelfth dynasty. We live in an age of Imperialistic ideas, and the Lake Mœris of to-day will be a series of reservoirs at the sources of the Nile. Lake Tana, at the source of the Blue Nile, with an area of 3,000 square kilometres, and catchment basin of about 18,000 square kilometres, will be able to supply 6,000,000,000 cubic metres of water per annum. Lakes Victoria and Albert Nyanza, at the sources of the White Nile, with areas of 70,000 and 4,500 square kilometres respectively, will be able to supply 12,000,000,000 cubic metres per annum. Lakes Tana and Victoria, with rocky sills and overfalls at their outlets, will not need to have their surfaces raised. The water will be drawn from them by widening the sills or cutting them down or tunnelling. Lake Albert will need to have its surface raised 3 or 4 metres by a weir at its outlet, but this will be done easily, as its shores are perfectly barren and within British territory. These mighty reservoirs will worthily fulfil their mission, a mission which has been thus described in the report of 1894 on the Nile Valley reservoirs: "The day these works are carried out at the sources of the Nile, the lakes will take their proper place in the economy of the water supply, and we shall be able to say of them in their entirety, as we can say of them to-day in their degree, that what the snows of the Alps are to the Po, Lakes Victoria, Nyanza and Tana are to the Nile; and that what the Italian lakes are to the plains of Lombardy, Lake Albert is to the land of Egypt."

In addition to these works at the sources of the Nile, training works will be undertaken in the White Nile itself between Gondokoro and Fashoda. These works will be undertaken with the object of ensuring that the waters issuing from Lake Albert are not being wasted in the marshes of the sudd regions, but are being conducted on to Egypt. They will not only add to the quantity of the water entering Egypt in summer, but also to its quality. Dr. Schweinfurth, the eminent African traveller and savant, was the first to call the attention of the Egyptian Government to the necessity of closing the spills from the White Nile to the north of Gondokoro, and so beginning the training of the river. He very rightly said: "Many years would elapse before the desired result would be obtained by the strengthening of the banks, but the works would be increasingly felt every year in Egypt as the works progressed."

When I wrote thus I had not realised the possibility of utilising the Assouan and Wady Rayan reservoirs in conjunction. It was only on the first anniversary of the opening of the Assouan dam that walking in the Fayoum within sight of the hills surrounding the Wady Rayan and regretting to myself that the ancient Lake Mœris could not be resuscitated, the idea of using the reservoirs together first dawned upon me. It was to me the morning of a new day.

To the solution of the problems connected with the sources of the Nile Sir William Garstin has devoted much time and study, and has twice visited these distant lakes and watery regions. In Egypt, however, we possess the means of storing all the water which is needed to provide perennial irrigation to the whole country on a truly liberal scale. The great advantage of having these two storage works situated within Egyptian territory is due to the fact that though it would be impossible for any hostile power in possession of the Sudan to shut off the flood waters of the great Abyssinian rivers, it would need no great engineering feat to shut off for a month or six weeks the low summer supply of the river. In case of such a contingency the Egyptian reservoirs would be of inestimable value to Egypt, the Equatorial reservoirs would be of no value whatever. Indeed, the existence of powerful regulating works and earthen embankments on the Nile in those distant regions might even be a source of danger, as has frequently been suggested.

I have, I think, fully kept the promise I made to you at the beginning of this lecture. I have shown you that the provision of perennial irrigation to the whole of Egypt is well within the competence of the finances of the country; and that it is moreover a work which presents no physical difficulties. That Egypt could easily afford to spend £2,500,000 on completing the irrigation system of the country is well proved by the fact that the country has just lent £2,000,000 to the Sudan to complete the Suakim-Berber railway. As far as the true interests of Egypt are concerned, however necessary the Suakim-Berber railway may be, the provision of perennial irrigation to the whole of Egypt is a hundredfold more necessary. No one has recognised this fact more fully than Lord Cromer, than whom irrigation has no truer friend in the world. The works which I have advocated could be begun immediately. Philæ temple stood in the way of the raising of the Assouan dam, and the inability to raise the dam stood in the way of the Lake Mœris project.

A great deal of the opposition to the annual inundation of Philæ temple for four or five months per annum came from a misapprehension. People confused the damage done by the salty percolation water, which filters up through the soil in flood time and works such havoc at Thebes, with the really preserving effect of the freshly flowing Nile water. One destroys stone, the other conserves it. The retaining walls of Philæ temple, which are annually inundated for 6 months per annum, are better preserved than any other part of the temple. It is the same with all stone and brickwork on the Nile. The submerged parts are better preserved than the parts which are never submerged.

M. Winston Churchill has well expressed the sentiments of the public on the lowering of the dam to save Philæ temple from an annual immersion which could have done it no possible harm. He has apostrophised this offering of 1,500 millions of cubic metres of water to Hathor by the wise men of the West, as—

> the most cruel, the most wicked, and the most senseless sacrifice ever offered on the altar of a false religion. The State must struggle and the people starve, in order that professors may exult and tourists find some place on which to scratch their names.

It might be possible to meet the wishes of both archeologists and tourists, and the requirement of Egyptian agriculture. As proposed by Sir William Garstin, it would be possible for the Department of Antiquities to remove Philæ temple on to Bigeh island. The whole of the dry stonework amounts to only 14,300 cubic metres. Sir William Garstin suggested handing over £200,000 to the Antiquities Department for this work, and then added:

> Prompted by a desire to benefit the country, we suggested the removal of an ancient building from one site of the Nile to another which is but a few hundred yards distant. We propose re-erecting it exactly as it stands to-day, and on an island in the middle of the great lake which we hope to create, where it would form a beautiful and appropriate object in the landscape.

The temple as it stands to-day, half submerged in water, is a sad object. Whenever I see it I imagine that Hathor is standing with wet feet and bedraggled garments beseeching to be removed on to Bigeh island. If the Department of Antiquities were to accept the £200,000, they could easily reconstruct the temple for half that sum, greatly improve its appearance by placing it as much higher than the full reservoir level as the temple was above the old Nile flood, and then spend the remaining £100,000 in thoroughly restoring Thebes and Luxor, Karnak and Abydos. Such opportunities come seldom to the Department.

We have seen how the whole of Egypt can be gifted with perennial irrigation and the value of the land raised from £275,000,000 to £335,000,000. We now come to the question of insuring for this future wealth an immunity from the terrible evils of inundation. In considering the question we must at the very outset either stand by the proverb which says: "Don't put all your eggs into one basket," or by the American modification which says: "Put all your eggs into one basket and watch that basket." I stand by the American modification.

Just as the design of the American engineers of the Nicaragua Ship Canal was chosen by me for the lock-gates of the Assouan dam, so now I propose for the training of the Nile the method recommended by Mr. Eads for the Mississippi. Mr. Eads's wonderful work on the Mississippi at New Orleans has justly won for him the reputation of being one of the soundest authorities on river training. With these two preliminary statements we may proceed to consider the best method of insuring Egypt against the evils of inundation.

The Nile during high floods is considerably above the level of the country, which is protected by embankments stretching from Assouan to the sea. In Upper Egypt a very high flood is one metre above the country, in Middle Egypt 2 metres, and the same on the Rosetta branch of the Nile; it is 3½ metres in places on the Damietta branch. The Damietta branch is really a canal and quite unfit to be used in high flood. The wisest policy, I think, would be for the Government to take the Rosetta branch and call it the Nile, and regulate on the Damietta Barrage just as though the Damietta branch were an ordinary canal. When we first came to Egypt we found that the policy was to spread the flood into as many channels as possible and protect the whole of them with tens of thousands of corvée. We changed that and concentrated our energies on the Rosetta and Damietta branches. Now that we can regulate on the head of the Damietta branch in flood, thanks to Sir Hanbury Brown's initiative, we might send the main flood down the Rosetta branch, and watch that branch. We should put our eggs into one basket and watch that basket.

In 1861, 1863, 1866, 1869, 1874, and 1878 the Damietta branch was badly breached. There has been only one serious breach on the Rosetta branch, and that was in 1863. The great breach of 1878 on the Damietta branch was attended with serious loss of life; but far more serious was the breach of 1863 on the Rosetta branch not far from its head. The whole western half of the Delta proper was swept by the river, and as the canals there have not got good high banks, the people had no place of shelter to flee to and were drowned in very great numbers. The same thing would happen again if a breach were to occur now, only the damage would be far more serious. The country is covered with villas and rich plantations and the low lands to the very edges of Lake Borillos are reclaimed and inhabited. The loss of life which would occur nowadays would be truly appalling. A breach anywhere within 100 kilometres of the Barrage on the east bank of the Rosetta branch or the west bank of the Damietta branch during a very high flood would be a national disaster.

The terror reigning over the whole country during a very high flood is very striking. The Nile banks are covered with booths at intervals of 50 metres. Each booth has two watchmen, and lamps are kept burning all night. Every dangerous spot has a gang of 50 or 100 special men. The Nile is covered with steamers and boats carrying sacks, stakes, and stone; while the banks along nearly their entire length are protected by stakes supporting cotton and Indian corn stalks, keeping the waves off the loose earth of the banks. In a settlement of a culvert in the Nile bank north of Mansourah in 1887 I witnessed a scene which must have once been more common than it is to-day. The news that the bank had breached spread fast through the village. The villagers rushed out on to the banks with their children, their cattle, and everything they possessed. The confusion was indescribable. A narrow bank covered with buffaloes, children, poultry, and household furniture. The women assembled round the local saint's tomb, beating their breasts, kissing the tomb, and uttering loud cries, and every five minutes a gang of men running into the crowd and carrying off the first thing they could lay hands on wherewith to close the breach. The fellaheen meanwhile, in a steady, business-like manner, plunged into the breach, stood shoulder to shoulder across the escaping water, and with the aid of torn-off doors and windows and Indian corn stalks, closed the breach. They were only just in time. This is the way the fellaheen faced a breach. And this is how the old Governors of Egypt faced them. During the flood of 1887 I complimented an official on the Nile bank, whose activity was quite disproportionate to his apparent age. He told me that he was a comparatively young man, but he had had charge of the Nile bank at Mit Badr when the great breach occurred in 1878, and that Ismail Pasha had telegraphed orders to throw him and the engineer into the breach. He was given 12 hours' grace by the local chief, and during that interval his hair had become white; subsequently he was pardoned. These were the senseless orders which used to petrify officials into stupidity.

The high level Wady Rayan Reservoir, when converted into the modern Lake Mœris, will have one great advantage, it will be able to lower a high flood 30 centimetres for 50 days. This will give relief to the Nile, a relief which will be much appreciated by the whole country from Beni Souef to the sea, and by Cairo.

I have already said that I should regulate on the Damietta branch and treat that branch like a canal. The whole of the energies of Lower Egypt could then be concentrated on the Rosetta branch, which is very capacious and capable of great development. The frequent breaches on the Damietta branch have been, among other causes, a great cause of the silting up of its bed and sides. The Rosetta branch has had only one real breach in 50 years.

Mr. Eads's argument is very clear. He insists that rivers eat away their banks in places, not owing to the direct action of the water but by the alterations in the velocity of the current. When the river water is charged with sediment to its full carrying capacity it cannot take up more unless the rate of current be increased. If the channel be nearly uniform the river water cannot eat away any of its banks. If, however, the channel is varying, the silt deposits in the wide sections, and the water, free of some of its sediment, is ready to eat more. It is this alternate dropping silt and eating away of earth which does the harm. To treat the Rosetta branch according to Mr. Ead it would be necessary to fix the top width to be worked to, at say 550 metres. The river could be brought to this uniform top width by building light inexpensive permeable spurs on the sandy shoals. The land between the spurs would become cultivated and such river training would pay the Government, which taxes all cultivated land; it would even pay handsomely for any company to undertake the work once the rule about foreshores was understood. The Government, would, however, always succeed; when it could not sell, it could always tax. Such training would permanently lower the flood.

In addition to the above it would be necessary to complete the system of spurs begun in 1884, and to throw back the banks as recommended on page 293 of my book on Egyptian Irrigation. Such banks might be thrown back 50 metres with advantage. It has been estimated that the completion of this work on the Rosetta branch, considering it the future Nile, would amount to £600,000; while the works on the Damietta branch, considering it a large canal, would amount to £300,000. The whole Delta would therefore cost £900,000 for spurs and banks, while the training works would pay for themselves in addition to greatly improving the channel and lowering the level of the flood.

In Upper Egypt there are 50,000 acres of sandy foreshore to be reclaimed and made worth £40 per acre by Eads's method of training, whose reclamation should be undertaken and which it would pay the Government to do. In the works of spurring and throwing back the bank, the cost will be greatly diminished in Upper Egypt by the long stretches of river where the desert comes to the water's edge and by the fact that stone in Upper Egypt is cheap. A sum equal to that for Lower Egypt would be amply sufficient to thoroughly train the Nile south of Cairo.

The works in Upper Egypt are not nearly so urgent as those in Lower Egypt, which should not be delayed on the Rosetta branch for a single year. Land in the Delta is worth 20 years' purchase to-day because there has been no breach since 1878 and irrigation and drainage are secured. How insecure the country is against a very high flood is known only to those who are responsible for guarding the banks. The utter absence of high floods for 9 years has lulled the country into a sleep from which there may be a rude awakening.

This river training by spurs is no new work in Egypt. The reach of the river between Assouan and Halfa is trained by gigantic stone spurs on the both banks. These works train the river in summer and aid navigation. They were probably put up by the great Rameses more than 3,000 years ago, as some of the most massive of them have evidently been constructed to turn the river on a curve out of its natural channel on to the opposite side in order to secure deep water in front of the temple of Jerf Hussin, a temple constructed by Rameses. These spurs have been constructed with great care and skill, and may even date back to the XIIth dynasty, as they were made by engineers.

It may be humiliating to make the confession, but from B. C. 2,200 to the Arab invasion of Egypt in A. D. 640, while Lake Mœris performed its allotted task and the Nile possessed training works such as those we can see to-day in Nubia, Egypt was better protected from inundation, and the Nile better trained, than it is to-day. And yet we have many advantages which no Pharaoh possessed. By the aid of telegraphy we have knowledge of a coming flood a full fifteen days before it arrives in the Delta; the Khartum gauge allows us to anticipate its very height. Meteorology is aiding us still further. In the paper I read at the Chicago International Exhibition I stated that years of heavy rainfall in India are years of high flood in Egypt, while years of poor rainfall in India are years of low flood in Egypt. Sir John Eliot, the Director General of the Meteorological Department of India, corrected this statement. He said that though this was not true of the Bengal monsoon, it was true of the Bombay monsoon. Years of heavy rainfall in Gujerat and Bombay are years of high flood on the Nile, and *vice versa*. As the rain falls in Bombay a month earlier than the Nile flood reaches Cairo, we have information of a high flood a month before it arrives, if we receive telegraphic information from Bombay. But I hope to go beyond this. I hope to establish that years of poor winter rainfall in the Levant and Mesopotamia, accompanied by weak Khamsin winds in Egypt, and followed by dry weather in the middle of June in the Nile Valley, foreshadow a poor flood; while on the contrary, years of heavy winter rainfall in the Levant and Mesopotamia, accompanied by strong Khamsin winds in Egypt, and followed by damp weather in the middle of June in the Nile Valley, foreshadow a high flood. We have meteorological information about Beyrout, Bagdad, Karachi, Bombay, Aden, Adis, Abeba, Khartum, Cairo and scores of intermedite stations, and with the aid of these we may be able not only to anticipate a high flood by fifteen or thirty days but by sixty days, and so find our hands greatly strengthened in grappling with the dangers of an inundation.

I bring to a close this lecture on the perennial irrigation and flood protection of Egypt by reminding you that I have purposely refrained from confusing your view with matters which are small indeed compared with the great issues we have been considering. These large issues I have attempted to face in all their rigour. The smaller issues have been kept in the background, not because they have been neglected or forgotten nor because they are unworthy of consideration. It has been done with the object of keeping your perspective clear. No criticisms have been indulged in on the engineering questions connected with the alignment of the dam or the level of the under sluices. These are small compared with the question of the stability of the dam and its ability to be raised. Similarly I have not referred to the question of the waters of the future lake eventually becoming salted and leaking through the limestone into the Fayoum. When the old Lake Mœris, or the present Fayoum, was full of water and 63 metres higher than the bottom of the Wady Rayan, and remained so for thousands of years, there was no question of the waters having become salted or having escaped into the Wady. The Wady was as dry as it is to-day and the great inland sea was always fresh. These questions were thoroughly threshed out at a meeting of this society in March, 1888, when Col. Ross strongly upheld the theory that the lake could not become salted. This position was confirmed by Dr. Schweinfurth, in an interesting communication to the Egyptian Government, a communication which was printed as an appendix to the Report of 1894. As to the question of the leakage into the Garak depression, a depression well nigh destroyed with soakage through sand, and one of the most desolate looking places in Egypt; it would be a Godsend to the place if more water were to leak into it, and the Government were to take up the question and pump this water into the Nezleh canal and utilise it in the irrigation of a few out of the tens of thousands of acres of desert to the west of the Fayoum which are awaiting water to become worth £20 per acre. Unfortunately no water will leak into it.

The great fact before us is the complete disposal within 12 months of the first milliard of cubic metres of water stored in the Assouan reservoir, and the eager demand for more from every quarter of the country. If the Government had at its disposal to-day the 3 milliards of cubic metres of water which are still required, they would be disposed of within an incredibly short space of time. The area which could be put under cotton every year in the country, provided there was sufficient water available in summer, would be 2½ millions acres. This area would insure 10 million cwt. of ginned cotton every year, as against an average of 6 million cwt. of ginned cotton to-day. Every acre of reclaimed land in Lower Egypt is specially suited to cotton. Such being the case, and the value of the water being so exceedingly great, and its storage so exceedingly easy, there is no reason why this storage should not be immediately undertaken. While I saw no way of utilising the Wady Rayan on its original lines, no one was more keen than myself in urging on the country the necessity of bringing this water from the equatorial lakes if it could not be obtained from elsewhere. Now, however, that the possibility of utilising this lake in conjunction with the Assouan Reservoir has dawned upon me I feel that everyone interested in the increase of the value of his land should encourage and call upon the Government to undertake both these works: the raising of the dam and the construction of the lake. Clearly as I saw the great advantage to the country of a liberal provision of perennial irrigation when I was Director General of Reservoirs and a member of the Irrigation Department of the country, now that I am Managing Director of a land and irrigation company, the advantages are seen tenfold more clearly. I do not only realise a rise in taxes from £1 to 1½ per acre, but a rise in rents from £3 to 6 per acre. This for Upper Egypt. In Lower Egypt the value of reclamation by the system introduced by Mr. Lang Anderson of shallow basins and drains 50 metres apart, into which the water percolates through the soil taking the salts with it, has revolutionised land reclamation and lifted perennial irrigation into a far higher plane than it even occupied in the past. There is no land in the Delta, however salted it may be, which will not yield to this system of drainage if perennial irrigation is provided.

The Irrigation Department is taking advantage of this fact to make drains much more effective than they could ever have been under the old wasteful system of surface washing. For this great work of reclamation, perennial irrigation is absolutely necessary. From whichever point of view we contemplate the situation the duty of water storage is as incumbent on us to-day as was the storage of corn on Pharaoh in Joseph's time. There is no task before the country, except it be the collateral one of protecting Lower Egypt from inundation, which can compare with this in magnitude; and it is to me no small satisfaction that, after 4 years' work in the field and 2 years' study in the office, followed by 8 years' persistent effort to solve the question, I have been privileged to lay before you a scheme which is as sure of realisation as it is certain to be remunerative, and of which, moreover, I can say with assurance that the world to-day possesses no project more full of hope.

Four thousand years ago, Egypt stood at the parting of the ways and adopted a system of water storage and flood control suited to basin irrigation, which served the country well for thousands of years. To-day Egypt stands again at the parting of the ways, and may it be her destiny to adopt a system of water storage and flood control suited to perennial irrigation which may stand her well for the thousands of years which may yet have to come.

Made in the USA
Monee, IL
04 November 2026

67388671R00017